Appreciation

Raymond E. Smith

Published by

Sanway International
91 E. Main Street
Inman, SC 29349

resco6042@yahoo.com

Website: www.SanwayInternational.com

Copyright © 2014 Raymond E. Smith

All rights reserved. No part of this book may be reproduced in any form or by any means without written permission from the publisher.

Dedicated to

Laurie Lantrip Delk for her constant remembering special days to show her appreciation to others.

Laurie is in the business of helping others to show appreciation. You can contact her at: www.ILoveCards.net or admin@LaurieDelk.net or 931.380.8811

Introduction

"You Needed Me" is a song written by Randy Goodrum, who describes it as being about "unconditional undeserved love." It was a number one single in the United States in 1978 for Canadian singer Anne Murray.

It was also recorded by Kenny Rogers and Dottie West

We all need to be loved and appreciated. Hopefully this book will lead you to having successful relationships by showing appreciation.

PART ONE

CHAPTER ONE

Showing Appreciation

In the beginning let me acknowledge that I am not the best example of showing appreciation. I make an effort by saying "Thank You," but like someone told me, "That's just words." Sometimes it takes more than words to show genuine appreciation. How to show appreciation will be discussed later.

Definition of appreciation according to Bing's Dictionary:

Gratefulness: a feeling or expression of

Appreciation

gratitude

Positive opinion: a favorable opinion of something

Valuing something highly: recognition and liking of something's qualities

Luke 17:12 *And as he entered into a certain village, there met him ten men that were lepers, which stood afar off: 13 And they lifted up their voices, and said, Jesus, Master, have mercy on us. 14 And when he saw them, he said unto them, Go shew yourselves unto the priests. And it came to pass, that, as they went, they were cleansed. 15 And one of them, when he saw that he was healed, turned back, and with a loud voice glorified God, 16 And fell down on his face at his feet, giving him thanks: and he was a Samaritan. 17 And Jesus answering said, Were there not ten cleansed? But where are the nine? 18 There are not found that returned to give glory to God, save this stranger. 19 And he said unto him, Arise, go thy*

Appreciation

way: thy faith hath made thee whole.

The response to showing appreciation to Jesus for what He had done may be pretty typical to the ratio we get today.

Leprosy is a horrible disease. Countries like India still deal with this disease today. In our support for orphans in different countries we encounter treatment for lepers.

In the above scripture reading we find Jesus healed ten lepers. He told them to go and show themselves to the Priest, which was the custom. When the Priest declared them whole, nine went their way and only one showed his appreciation by worshipping Jesus.

In 2006, after hurricane Katrina hit the gulf coast of the United States, New Orleans, Louisiana was one of the hardest hit areas. A non-profit group of which I was president wanted to offer our help.

Toby McCall, a friend from church, and I

Appreciation

drove down to survey the damage and see where we could be of service. We made the 1,500 mile round trip and met with an Elder of a church. This church had taken on a large responsibility of helping people in the city with personal needs as well as their homes and churches.

We could see right away that the church had accepted more than they could handle. Contact was being made with other churches and individuals around the country making a plea for help.

Returning home we started to organize a work crew to make a return trip. A total of nine men volunteered to go and work.

The Central Church of Christ in Spartanburg, South Carolina organized a group to fill personal care packets. A gallon-size plastic zip-lock bag was used to hold items such as tooth paste, tooth brush, hand soap, wash cloth and other personal items. Several hundred bags were filled. We also loaded a trailer with other useful items.

Appreciation

Arriving in New Orleans, tired from the long trip, we unloaded the trailer and went to work immediately. While some of the crew broke up a concrete floor with jack hammers, the other went to get our assignment for the next few days.

The work group consisted of nine men, which included Toby McCall, Tommy McCall, Al Smith, Tim Fagan, Dave Long Jon Hall, Jeff Perry, Robert Dickerson and me.

We were sent to another part of the city, actually another Parish. Our job was to go to a school building that had been flooded by the hurricane and build 16 shower stalls. Workers could camp on the second level but the shower needed to be on the first level.

First the electricity has been damaged and shut off the entire building. Two rooms had to be built to separate the men's and ladies showers. God provided us with the right guys to handle the work. Toby, a plumbing contractor and his assistant Al

Appreciation

Smith handled all the plumbing detail. Tommy, an electrical contractor and Robert Dickerson handled the electric work, which was a tremendous job. As far as the construction of the rooms, I have about sixty years experience in construction. The rest of the guys and I did the carpentry work. One or two of the guys spent endless hours at the nearest Home Depot buying materials.

With a multitude of set-backs, disappointments and unforeseen responsibilities, we finish the job in eight days and some nights.

Standing back looking at the job and proud of what we accomplished, I for one keep waiting for the Elder who was in charge to come by and see the good job we did. Since he did come I thought maybe he will call to offer his appreciation. Since that didn't happen, after getting back home I waited for a letter or maybe just an email to express his appreciation. That didn't come either. A non-profit organization needs letters of appreciation in their

Appreciation

files. We lost contact with them.

The following years we discussed whether or not we should go back to New Orleans. Feeling somewhat unappreciated we decided to raise money to build an orphans home in Zambia, Central Africa.

Just to emphasis the point, if the Elder in New Orleans had shown some appreciation we would have been more than glad to make another trip.

CHAPTER TWO

Who Needs Appreciation?

The bottom line is everyone needs to be appreciated. In this chapter we want to break it down and see some categories of individuals who need appreciation.

According to John Groberg our ego has a lot of needs. It wants to be safe, it wants to be comfortable, it wants to be stimulated. But most of all it just wants to be appreciated. Appreciation is the deepest need of our ego- deeper even than love (at least what the ego thinks of as love). The ego

Appreciation

craves appreciation and it will go to great lengths to find it from outside itself. If it isn't feeling appreciated in a job or in a relationship, it will soon start seeking a new job or a new relationship where it thinks it can find more appreciation. We have the power to give our ego the appreciation it craves, but all too often we think that appreciation is something that comes from outside us - from others. We neglect to appreciate our own egos- often because we've been taught and believed that our egos are part of our lower nature and need to be overcome and subdued. On the other end of the spectrum, we may indulge in false flattery of the ego, trying to make ourselves feel important and better than others or desperately seeking the approval and appreciation of others and willing to overlook our own standards in order to get it. I believe that a lack of appreciation for the ego combined with a lack of leadership of the ego by the Spirit is the root of most of our suffering and unhappiness.

Learn more about John Groberg from his

Appreciation

website http://www.johngroberg.com.

The deepest principle in human nature is the craving to be appreciated." - William James

"There is more hunger for love and appreciation in this world than for bread." - Mother Teresa

Business Associates

I believe there is a way to grow business relationships by simply using words of encouragement! There is nothing that will get me to work harder than words from my employers or co-workers that I'm doing great or keep up the good work. Just as when I'm constantly being put down by my employers my quality of work and focus will slip. I think it is very important that to have a good healthy work relationship you need have words of encouragement and healthy pointers on how to improve in your work. On a side note I do love gift certificates for a pedicure!!! - Jolene

Appreciation

Sopalski

What is the best gift you ever got from a business associate? Here are the answers: Pens, Monte Blanc pen, gold ink pen, emery boards, a nice bottle of wine, calendar, clock, sash chocolate, cookies, coffee mugs, design poster signed by the artist, dinner for two, food for the staff, tape measure, gift basket, gift certificate, leather checkbook cover, leather coasters, letter opener, screw driver, Christmas poinsettia, small flashlight, smile stamp, thank you card and watch! – Toni Blake

My attorney settled an insurance claim for me. At the settlement when he presented the check, he also has me a bag of delicious home-made oatmeal cookies.

Employees Need Appreciation

People respond positively when treated kindly. Employees work harder when they feel appreci-

Appreciation

ated. With a sincere, well-written appreciation letter you can make someone feel special or improve relations. Focus on words of admiration, gratitude and recognition.

Write your letter quickly. Write your appreciation letter within a few days of whatever happened to inspire your gratitude.

Use stationery. Write your personal appreciation letter on a sheet of stationery.

Use customized letterhead for business correspondence.

Address your appreciation letter to a specific person. If possible, not just to the company or to the organization in general.

Specifically state what you appreciate.

Be sincere. Most people can sense when you are not being honest.

Write clearly and concisely. This is no time to be longwinded or flowery.

Appreciation

Be brief.

Avoid thanking the person beforehand. For example: "Thanking you in advance for your help in this matter." To do so is presumptuous and suggests you do not feel the need to write a follow-up letter. Close with warm regards or best wishes for continued success.

Meet Your Husband's Need for Appreciation

A man needs to feel appreciated for the good things he does. If he doesn't feel appreciated, he may stop doing those things. You, as his wife, are the one who can best meet his need to feel appreciated for most of the things he does. A man needs to feel appreciated for the paycheck he brings home. Every household has to make decisions about money and how it will be spent. These decisions can be stressful. Try to keep the stress separate from his

Appreciation

ability as a provider, rather than blaming him for the difficulty. Say it, and show it

Your feelings of appreciation for your husband may motivate you to do the grocery shopping, scrub the floors, and work long hours to help pay the bills. Although they are important expressions of appreciation, such mundane acts do not generally communicate appreciation to your man's brain with sufficient force to fuel romantic love, perhaps because their motives can be variously interpreted (Maybe you just like the look of a freshly scrubbed floor.) Rather, appreciation for him must be clearly stated with frequent verbal communication "I really appreciate you facing the dragons each day at work" or "Thanks for taking out the trash."

Men feel less of a need for little gifts or written expressions of appreciation than women do. They are more likely to understand and respond well to visual and physical affection as expressions of appreciation. Show him that you appreciate spe-

Appreciation

cific things he does by making yourself especially attractive or giving him some extra sexual attention.

Wives Needs Appreciation

A woman needs to feel appreciated for the good things she does. If she doesn't, she may cease doing those things. You, as her husband, are the one who can best meet this need. Tell her frequently how much you appreciate the little and big things she does. Compliment her on her intelligence, abilities, and appearance, and tell her how much you admire her. Be sure to show your gratitude for her contributions to the financial wellbeing of the household, whether

it is bringing home a paycheck or cooking more at home and budgeting well. Your frequent expressions of appreciation will help her to gain a greater sense of self worth and confidence. As she does, she will become a more valuable, admirable, beautiful, and praiseworthy woman.

Appreciation

Say it with words. A man feels love through his eyes, a woman through her ears. Your woman needs to hear your expressions of appreciation for what she does: "The spaghetti sauce you made is delicious." "Thanks for making dinner." I know a lady who was separated from her husband. I asked her to tell me what the problems were. She specifically said, "He never tells me he enjoyed the meal I cooked." "You sure are a good mother to our children" "It was so nice of you to ask my opinion." These words can be made more meaningful if accompanied by a token gift, such as flowers or jewelry.

Express your appreciation in words both written and spoken. Although you may not need to receive cards from her on special occasions, she does need to receive them from you. Send her a card for her birthday, your anniversary, Mother's Day, and Valentine's Day at a minimum, and on other days occasionally "just because" you love and appreciate her. Alternatively, give her flowers

Appreciation

with a written note expressing your love. A handwritten letter mailed to her will also turn her heart.

You may also effectively convey your appreciation by acts that are so bizarre that no alternative motive can be assigned to them (for example, unexpectedly washing her car and leaving a note on the steering wheel expressing your appreciation for something she has done).

Setting aside time for your spouse is critical for the outworking of a successful family. A strong marriage bond can withstand any trial that may face the family as a whole. The trial may be as minuscule as a late bill payment or as colossal as the loss of a job. Whatever the case may be, having two strong heads of the family will provide you with security and the satisfaction of knowing that you can get through whatever comes your way.

Showing Parents Appreciation

Appreciation

Parents are often taken for granted by their kids. If you and your kids get along great, then they probably don't realize that a thank you to a parent is important. You need to sit down and have a serious talk about this with them. If they also ignore other people who give them gifts, then tell them that their lack of appreciation makes them appear selfish and ungrateful. People judge others by their actions. Explain how their lack of appreciation makes you feel (hurt, sad, unappreciated, etc.). Let them know that there will be a consequence to their not appreciating the gifts that you give. Decide what the consequence will be before you have the talk. For example, you will give a monetary contribution to a charity in their name instead of a direct gift to them. It's important that you follow through and really do something other than give them a gift. Always call or send a card to let them know you're thinking about them, and let them know about the donation in their name.

Appreciation

Showing Children Appreciation

According to Vanessa Brown in her story, "I Never Grew Up," "I sometimes find myself being too negative with my children. I will hear myself correct them, ask them to stop things, hurry and start things way more often than focusing on positive things. I will tell them "Good Job" on different tasks like cleaning up their room or finishing the homework.

But I realized that I didn't give them genuine "thank yous" enough. Which made me a bit sad to realize, but now I have quickly gotten in the habit of showing appreciation to my children more often. Showing appreciation to our children and also teaching them to show appreciation to each other can be done in many fun and playful ways."

Remember children don't always want things. The best way to show them appreciation may be to spend quality time with them. Go to your children's sports activities. You can see the hurt in

Appreciation

the child's eyes when he says, "My daddy never comes to see me play ball." Sometimes better than watching them play would be to play ball with them.

When did you last take a walk with your daughter? Or, climb up the ladder to see the inside of the tree house?

Showing Teachers Appreciation

Showing teacher appreciation to your children's educators is a good investment in your school system. It also is a great way to get to know the teachers more and work with them to provide your child with the best possible academic experience. As opposed to grumbling and complaining, a teacher thank you gesture goes a long way to help teachers feel valued and appreciated. Some schools celebrate a teacher appreciation day with hot meals for lunch, offer special gifts or host a classroom party to honor their children's teachers. If your

Appreciation

kids' school does not practice some form of collective teacher appreciation, why not take the initiative to tell your kids' teacher thank you with a thoughtful expression of appreciation?

Showing Appreciation to Your Minister

Why is minister appreciation so important? Because a pastor that is appreciated by the majority in his congregation is a minister that can press forward confidently knowing that his people are with him.

Also because he knows that he is not carrying the burdens of ministry alone. What ministers need most is people who will partner with them, have a vision for the ministry of the church, and roll up their sleeves and get involved in the real (and sometimes messy) ministry of encouraging, discipline, leading, and teaching.

When Jesus gave the great commission in

Appreciation

Matthew 28:19 "Go ye therefore, and teach all nations, baptizing them in the name of the Father, and of the Son, and of the Holy Ghost: 20 Teaching them to observe all things whatsoever I have commanded you: and, lo, I am with you alway, even unto the end of the world," He didn't give this to a group of ministers, He gave it to individuals.

Jesus was also talking to individuals rather than a group of ministers when He said in Matthew 25:35 For I was an hungred, and ye gave me meat: I was thirsty, and ye gave me drink: I was a stranger, and ye took me in: 36 Naked, and ye clothed me: I was sick, and ye visited me: I was in prison, and ye came unto me.

When you meet the minister at the door after a good sermon and say, "I enjoyed that," that may be just words. It sounds genuine and shows appreciation when you mention something he said that really interested you.

Appreciation

Showing Appreciation to Customers

Several years ago I moved to East Tennessee to start a construction business. Most of my initial contacts for doing home improvements were women. I hit on the idea of doing something to show my appreciation when a job was completed. I contacted a local florist to send a dozen long stem red roses at the completion of the job. The idea was a great success. I soon became one of the most popular contractors in town.

CHAPTER THREE
Ways of Showing Appreciation

Quality time: The older people become the more they need quality time. Spend quality time with elderly family members on a regular basis. In fact, make a recurring date on your calendar to take your great uncle out to lunch. Although you can't confuse quantity with quality, he may treasure a weekly lunch date more than a twice yearly fishing trip. Increasing the time you spend with an elderly person and setting a firm date gives the relative the message that he matters, and it fills his week with

Appreciation

anticipation.

    As parents we strive to support our families both financially and emotionally. It is easy to lose sight of our priorities when our careers demand so much of us. Granted, each one of us must make a living to support our families, but we must also take into consideration our family's emotional well-being. By setting aside time for our children as well as our spouse, we'll strengthen their beliefs that they are loved and they are an important part of the family.

    Gifts: People love to get gifts. They do not need to be expensive. As I mentioned earlier my attorney gives a bag of home-made cookies to his clients at the closing of a deal. Some companies specialize in sending various gifts to customers. It may be a box of brownies, flowers, and etc. Gifts do not always need to be expensive. In some cases a discount coupon for future use may be effective.

Appreciation

Acts of service: When we appreciate what someone has done for us, for someone else, the community, or the country do something for them. Various groups are being created to show our military personnel appreciation. Some are building houses, special welcome home, artificial limbs and many other ways. Find out where these groups are and get involved.

Words of affirmation: This is the simplest way to show appreciation. Not just words, but sincere words to affirm how you feel. "Thank you for helping me with my assignment." I appreciate you offering me a ride home when my car was broke down." "I appreciate you backing me up on my decision." "Thank you Mom for making the pie, you make the best." Think of ways you can show appreciation with words. Complete the following:

Appreciation

I appreciate you because:

… you come to work on time.

… you are always courteous.

… you have a nice smile.

… you keep your desk neat.

… you have a smile in your voice on the telephone.

… you care about others.

… your meals are always delicious.

… you get to the church services on time.

… you always bring in your homework.

… you make your payments before they are due.

… you take out the trash before being told.

… you keep your room so clean.

This list could go on and on. Showing appreciation with words becomes a habit and a natural response.

CHAPTER FOUR

Why Show Appreciation?

It is less expensive to show appreciation in order to keep a customer than to develop a new one.

In Dale Carnegie's "How to Win Friends and Influence People," one of the most important qualities he mentioned in getting along with people is honest and sincere appreciation. This has proven to be quite true in reality for me, from both the giving and receiving end. I have noticed people's eagerness to help out when little things they do are remembered, and also experienced firsthand the dev-

Appreciation

astatingly de-motivating effects when my efforts go unnoticed.

When you don't show appreciation is basically telling people, I don't care how you feel.

In his book, My Shadow Ran Fast, Bill Sands tells about getting into trouble with the wrong crowd. At age 12 he spent time in a correctional institute. A few years later he robbed a store and shot the owner. He had lost his mother and his father didn't have time for him.

In prison at San Quentin he was determined to be the worse prisoner there. He started fights just to be put into solitary confinement. The letters he sent to his father were returned. He became more and more bitter every day.

One day Warden Duffy went into his cell to talk with him. The warden encouraged him to improve so he could get out of "the hole." Bill's answer was, "Who cares?" With a heart full of compassion the warden said, "I care." That was the be-

Appreciation

ginning of Bill changing his life.

Over a period of time Bill became a model prisoner. He worked at Warden Duffy's home. In time he was paroled. Bill spent the rest of his life helping people and trying new things to fulfill his quest for living to the fullest.

Bill's life would most likely have never changed but someone said, "I care." Who can you tell, "I care?" This may be the most important gift of appreciation you'll ever present to anyone.

Making Appreciation Your Business

Cynthia Powell is an entrepreneur that found her niche in business, showing appreciation. She has a baby shoe bronzing business and she is the Manager of Sending Out Cards.

My Baby Shoe Bronzing Business

I do all my marketing for this business based on being in the top of Google for Baby Shoe Bronzing. When someone comes to my site and wants more information they have to subscribe to my email list. And for years that is how I have communicated with them.

In the fall of 2010, I started adding Thank You Cards and follow up phone calls to the mix. When they request information, I ask for address and phone number. I then follow up with a Thank You card, a phone call in 2 months and then anoth-

Appreciation

er phone call at 5 months.

In 3 months' time, I did as much business as I had done in the 6 months before. MULTIPLE ways of communication, be appreciative and helpful.

Customer Testimonial: "You are one of the best business people I have encountered in a long time. Your follow up is top notch. I will be passing your name along to friends and family."

Thank you . . . for telling your ABC Company story.

T, he business of a good friend of mine was affected by the economy. She will tell you that what SAVED her business – thank you cards with a Starbucks Gift card tucked in. Here is what she did:

The front of the card was THANK YOU. The inside left flap had a great customer testimoni-

Appreciation

al about her business. Inside Right flap said: "Thank you for telling your ABC Company Story." And, of course, the $5.00 Starbucks Gift Card. Her phone started ringing off the hook with referrals.

LOVE they can HOLD – read and re-read. "I came home from being away this weekend and had a thank you note in the mail, from my boss, thanking me for doing exemplary work. It made me feel like a million bucks. And today, I picked it up and read it again, and it made me feel like a million again! I have never been in a job where I felt so appreciated! It's a good feeling!" Think of the POWER in an "I'm Proud of you Card" to someone. It is important to SAY things but when it is put into writing, in a card; – it is more permanent to them.

Showing appreciation can sometimes be a sure way to mend a relationship. Think of all the ways you can show your friends, relatives and business associates appreciation. Start small and work your way to bigger things.

PART TWO

Dealing with Relationships

Preface

People have varying degrees of comfort with conflict. Some prefer avoiding it at all costs. Unfortunately, those costs tend to increase the longer issues are left unaddressed. Therefore, learning how to manage and resolve conflict is to your benefit. When dealing with conflict:

> Treat it as normal and expected. Conflict need not be catastrophic or

Appreciation

personal. Conflict is simply part of being human.

Deal with issues as they arise. Avoiding conflict makes situations worse. Time does not resolve matters. Instead, it decreases the chance of a positive outcome.

Attempt to understand the other person's point of view. Dismissing the other's views, assigning blame, and exclusive focus on your own perspective are all counterproductive.

Don't judge emotions. No one's feelings are more or less "right" than the others. Emotions reflect a valid perspective of an individual. Even if you don't understand it, acknowledge the other person's reaction as important.

Focus on the behavior, situation or

Appreciation

problem area without attacking the person involved.

Do not assume your values or beliefs are "right." They reflect a view of the world from your unique perspective. Respecting another's viewpoint as equally valuable opens an opportunity for learning and growth.

Dealing with conflict does not need to be dreaded or feared. Interpersonal conflict is a natural component of human interaction. In fact, if the problem is the object of focus versus the people involved, disagreements can generate new ideas and growth. Dealing with issues as they occur, acknowledging the other party's feelings and perspective, and avoiding judgment or blame further increase the chance of pr

Appreciation

productive conflict resolution.

CHAPTER ONE
Keep a Positive Attitude

We live in a mixed up, complex world. Sometime keeping a positive attitude is very difficult. With effort we can overcome the negatives that we are faced with on a daily basis.

The story was told about a brick mason who had finished building a very tall chimney. He realized that he had a lot bricks leftover. He had a dilemma; how would he get them down. If he throws them down, from that height, surely they would break.

Appreciation

He thought about his problem over night. The next morning he tied his rope to a 55 gallon steel drum. He pulled the rope across a pulley and tied one end to a tree. From the ground he pulled the drum to the top. Then he filled the drum with the bricks.

He untied the rope from the tree in order to let the drum down slowly. But, the drum with bricks weighed more than he did. He was afraid to turn lose of the rope, so he held tight. As the drum of bricks came down he went up. Meeting the drum half way he got a terrible lick on the head. The drum hit the ground so hard it knocked the bottom out and the bricks spilled on the ground. The drum is now lighter than his weight, so he starts coming down and the drum goes up. Again half way he gets another lick from the drum passing him. He hits the ground and loses control of the rope. That allowed the drum to fall uncontrollable and lands on top of him.

Appreciation

As you might expect the mason had major injuries. Plus he was highly aggravated with his plan to get the bricks down.

Sometime our day goes about as wild as the brick mason's. How do you stay positive under these circumstances?

Start by getting your attitude right:

Our attitude is something over which we have complete control. Let me give you an example. You have a bad day at work. On the way home you have a flat tire and ruin your freshly dry cleaned suit. You get home and during dinner you three year old spills a glass of milk and it goes all over you. About that time the telephone rings. With a few ugly words you answer it with an unpleasant H-E-l-l-O what do you want. Then, with a total change of voice and attitude, you say, "Oh hi Preacher I didn't know it was you."

Appreciation

An attitude can be changed as fast as snapping you fingers. All it takes is an attitude adjustment.

You can choose the right attitude. (Philippians 2:14) *"Do everything without complaining or arguing, so that you may become blameless and pure, children of God without fault in a crooked and depraved generation, in which you shine like stars in the universe as you hold out the word of life--in order that I may boast on the day of Christ that I did not run or labor for nothing."*

Attitudes are more important than facts. (1 Samuel 16: 6-7) *When they arrived, Samuel saw Eliab and thought, Surely the LORD's anointed stands here before the LORD."But the LORD said to Samuel, Do not consider his appearance or his height, for I have rejected him. The LORD does not look at the things man looks at. Man looks at the outward appearance, but the LORD looks at the*

Appreciation

heart."

We sometimes get out of alignment with God. (1 Samuel 13:13-14) "You acted foolishly," Samuel said. "You have not kept the command the LORD your God gave you; if you had, he would have established your kingdom over Israel for all time. But now your kingdom will not endure; the LORD has sought out a man after his own heart and appointed him leader of his people, because you have not kept the LORD's command."

When he sinned he prayed for forgiveness. (Psalms 51:1-10) *"Have mercy on me, O God, according to your unfailing love; according to your great compassion blot out my transgressions. Wash away all my iniquity and cleanse me from my sin. For I know my transgressions and my sin is always before me. Against you, you only, have I* sinned and done what is evil in your sight, so that you are proved right when you speak and justified when you judge. Surely I was sinful at birth, sinful from

Appreciation

the time my mother conceived me. Surely you desire truth in the inner parts; you teach me wisdom in the inmost place. Cleanse me with hyssop, and I will be clean; wash me, and I will be whiter than snow. Let me hear joy and gladness; let the bones you have crushed rejoice. Hide your face from my sins and blot out all my iniquity. Create in me a pure heart, O God, and renew a steadfast spirit within me."

Getting right with God is a good attitude adjustment. Our sins are against God.

Attitudes are sometimes caused by prejudging. Did you ever decide what a person would be like before you met them?

Someone makes a mistake at the office. "I've told you a thousand times…"

Peter walked on water as long as his attitude was right.

Appreciation

Expect results:

Let's look at a mother's attitude. (Mark 7: 24-30) Jesus left that place and went to the vicinity of Tyre. He entered a house and did not want anyone to know it; yet he could not keep his presence secret. In fact, as soon as she heard about him, a woman whose little daughter was possessed by an evil spirit came and fell at his feet. The woman was a Greek, born in Syrian Phoenicia. She begged Jesus to drive the demon out of her daughter. "First let the children eat all they want," he told her, "for it is not right to take the children's bread and toss it to their dogs." "Yes, Lord," she replied, "but even the dogs under the table eat the children's crumbs." Then he told her, "For such a reply, you may go; the demon has left your daughter." She went home and found her child lying on the bed, and the demon gone.

Appreciation

Be Forgiving:

The case of the two sons. (Luke 15) *"And he said, A certain man had two sons: 12 And the younger of them said to his father, Father, give me the portion of goods that falleth to me. And he divided unto them his living. 13 And not many days after the younger son gathered all together, and took his journey into a far country, and there wasted his substance with riotous living. 14 And when he had spent all, there arose a mighty famine in that land; and he began to be in want. 15 And he went and joined himself to a citizen of that country; and he sent him into his fields to feed swine. 16 And he would fain have filled his belly with the husks that the swine did eat: and no man gave unto him. 17 And when he came to himself, he said, How many hired servants of my father's have bread enough and to spare, and I perish with hunger! 18 I will arise and go to my father, and will say unto him, Father, I have sinned against heaven, and before thee, 19 And am no more worthy to be*

Appreciation

called thy son: make me as one of thy hired servants. 20 And he arose, and came to his father. But when he was yet a great way off, his father saw him, and had compassion, and ran, and fell on his neck, and kissed him. 21 And the son said unto him, Father, I have sinned against heaven, and in thy sight, and am no more worthy to be called thy son. 22 But the father said to his servants, Bring forth the best robe, and put it on him; and put a ring on his hand, and shoes on his feet: 23 And bring hither the fatted calf, and kill it; and let us eat, and be merry: 24 For this my son was dead, and is alive again; he was lost, and is found. And they began to be merry."

The boy wanted to be free. He might have been tired of hearing "Where did you go last night?" "Who were you with?" "What time did you get home?"

The father could have said, "I don't like your attitude." Instead he gave his son what he asked

Appreciation for.

Learn to be Content:

(Philippians 4:11) "I am not saying this because I am in need, for I have learned to be content whatever the circumstances."

Paul had been beaten, put in prison, stoned, bitten by a snake, ship wrecked and chased out of town many times. Yet he chose to keep a content attitude.

Analysis your attitude:

Too often our attitude is "Gotta win," "Gotta be consulted," or "Gotta be seen." Remember your attitude is something that you control.

CHAPTER TWO

Do It Whether You Want To or Not

"You can't control the wind, but you can adjust your sail." Do it because it is the right thing to do. In many of my seminars I encouraged people to memorize this affirmation and repeat it many times a day. "I will do the right thing because it is the right thing to do."

Joni Erikson Tada has been in a wheel chair more than 40 years because of a diving accident. She chooses her attitudes and goes to work every day. She travels and gives speeches. She cannot use

Appreciation

her arms yet she drives a van with voice command.

We can choose our attitude:

Turn your thoughts, attitudes and actions over to God. We have to say, "Father your thoughts are my thoughts, your ways are my ways, I'll do whatever you want me to do."

Total trust in God to handle our problems is like signing a contract and telling God to fill in the blanks.

William James, Harvard psychologist said, "The greatest discovery in our generation is that human beings can change their lives by changing their attitudes."

Happiness and satisfaction in life are not determined by our actual circumstances, but by our attitude toward those circumstances. Think about Paul's circumstances.

Being content depends on your ability to be

Appreciation

satisfied with what you have, and not what other people have. This happens in the work place, in schools, factories, on construction jobs and etc.

At this writing we are going through a very bad economy. Many people who are trained for higher jobs have to work for less wages. (Matthew 20: 1-15) *"For the kingdom of heaven is like a landowner who went out early in the morning to hire men to work in his vineyard. He agreed to pay them a denarius for the day and sent them into his vineyard. About the third hour he went out and saw others standing in the marketplace doing nothing. He told them, `You also go and work in my vineyard, and I will pay you whatever is right.' So they went. He went out again about the sixth hour and the ninth hour and did the same thing. About the eleventh hour he went out and found still others standing around. He asked them, `Why have you been standing here all day long doing nothing? Because no one has hired us,' they answered. He said to them, `You also go and work in my vineyard.'*

Appreciation

When evening came, the owner of the vineyard said to his foreman, `Call the workers and pay them their wages, beginning with the last ones hired and going on to the first.' The workers who were hired about the eleventh hour came and each received a denarius. So when those came who were hired first, they expected to receive more. But each one of them also received a denarius. When they received it, they began to grumble against the landowner. `These men who were hired last worked only one hour,' they said, `and you have made them equal to us who have borne the burden of the work and the heat of the day.' But he answered one of them, `Friend, I am not being unfair to you. Didn't you agree to work for a denarius? Take your pay and go. I want to give the man who was hired last the same as I gave you. Don't I have the right to do what I want with my own money? Or are you envious because I am generous?'

Some worked all day, some worked 1 hour and all got the same pay. The point is to do what

Appreciation

you agreed to do.

I have personally had a similar situation happen in the construction business. I had worked for a company four years. When they hired new superintendents they were paid a higher salary than I was getting. Was it fair? No. But, I had to be content with it.

Being content does not mean accepting whatever circumstances send you. Through training, education and personal experiences you can create better situation in your life.

How do you change your attitudes?

Believe you can change your circumstance with God's help. Paul learned to conquer circumstances. (Philippians 4: 11-13) *"I am not saying this because I am in need, for I have learned to be content whatever the circumstances. I know what it is to be in need, and I know what it is to have plenty. I*

Appreciation

have learned the secret of being content in any and every situation, whether well fed or hungry, whether living in plenty or in want. I can do everything through him who gives me strength.

Paul made no claim to be supernatural. Have a good role model. (Habakkuk 3:17-18*)* *"Though the fig tree does not bud and there are no grapes on the vines, though the olive crop fails and the fields produce no food, though there are no sheep in the pen and no cattle in the stalls, yet I will rejoice in the LORD, I will be joyful in God my Savior."* Man – what an attitude to have.

Choose a role model with great character. Try this for changing your attitude: "Though I am living in a recession, and have lost my income, my unemployment benefits have expired, I am hungry and out of groceries and don't know where my next meal is coming from, yet I will rejoice in the Lord."

The way to change your attitude is to believe

Appreciation

that through Christ, you can do it.

Being willing to change your attitude doesn't mean "giving in." You can will to, whether you want to or not. Abraham Lincoln said, "People are about as happy as they choose to be." Don't let your "will" control your "wants."

(John 7: 17) *"If anyone chooses to do God's will, he will find out whether my teaching comes from God or whether I speak on my own."*

(Philippians 2: 12-13) *"Therefore, my dear friends, as you have always obeyed--not only in my presence, but now much more in my absence continue to work out your salvation with fear and trembling, for it is God who works in you to will and to act according to his good purpose."*

Some are not willing to change. The prodigal son was willing to change. He said, *"I will arise and go to my father."* He made a decision to go whether he wanted to or not.

Appreciation

Change the way you think:

(2 Corinthians 10: 5) *"We demolish arguments and every pretension that sets itself up against the knowledge of God, and we take captive every thought to make it obedient to Christ."*

A 25 Year study at Harvard revealed, "You can change motivation and performance by changing the way you think."

Have a condescending attitude:

(Romans 12:16) *"Live in harmony with one another. Do not be proud, but be willing to associate with people of low position. Do not be conceited.* "King James version says, *"but condescend to men of low estate."*

Have the mind of Christ. How does Christ think? He came down from Heaven. He was placed in a feed trough for a bed. He worked as a carpenter's helper. He humbled himself.

Appreciation

(Philippians 2: 3-14) *"If you have any encouragement from being united with Christ, if any comfort from his love, if any fellowship with the Spirit, if any tenderness and compassion then make my joy complete by being like-minded, having the same love, being one in spirit and purpose. Do nothing out of selfish ambition or vain conceit, but in humility consider others better than yourselves. Each of you should look not only to your own interests, but also to the interests of others. Your attitude should be the same as that of Christ Jesus: Who, being in very nature God, did not consider equality with God something to be grasped, but made himself nothing, taking the very nature of a servant, being made in human likeness. And being found in appearance as a man, he humbled himself and became obedient to death-- even death on a cross! Therefore God exalted him to the highest place and gave him the name that is above every name, that at the name of Jesus every knee should bow, in heaven and on earth and under the earth,*

Appreciation

and every tongue confess that Jesus Christ is Lord, to the glory of God the Father. Therefore, my dear friends, as you have always obeyed--not only in my presence, but now much more in my absence-- continue to work out your salvation with fear and trembling, for it is God who works in you to will and to act according to his good purpose. Do everything without complaining or arguing."

Reconsider the way you think:

Maybe you are too close to the forest to see the trees. (James 1: 2-4*)* *"Consider it pure joy, my brothers, whenever you face trials of many kinds, because you know that the testing of your faith develops perseverance. Perseverance must finish its work so that you may be mature and complete, not lacking anything."*

Appreciation

Use it or lose it:

My arm was in a cast following surgery. Because of a staph infection it stayed in the cast a few more days. When the cast was removed I could not move my arm. I spent two more weeks in another hospital with trips back and forth to surgery trying to physically break the joint loose. Finally, with weeks of physical therapy I got 90 percent range of motion.

Stop acting in faith and faith dies. Stop hoping and hope dies. Stop loving and love dies.

In the Old Testament God required food brought for a thank offering. He didn't need the food. He knew they need to express their thanks through actions.

CHAPTER THREE
Act Better Than You Feel

Cognitive dissonance: that's an odd phrase. Cognitive means to believe. Dissonance means opposition. It's like eating a giant Baby Ruth candy bar with a Diet Coke.

There was a sign: "When you give someone a helping hand, you open the door to teach the gospel." Example: John Hazelip a preacher in Jacksonville, FL. Some men were remodeling the church building when suddenly a member ran in to John's office and said, "Quick give me some scriptures on

Appreciation

baptism." A little while later he came back and said, "Quick give me some scriptures on the Holy Spirit." John said, "What's going on." He told him there is a man they hired to do some painting and he wants to convert him before he leaves. John asked, "What do you know about this man?" He admitted, "Not much, he has three small children and they live in a small travel trailer." John said, "You are going about this all wrong. Bring the children in and wash their faces and feed them. Hook the trailer up to the electricity. Let them get some rest then we will teach them the Bible."

Our lives would be better if we always felt like doing what we ought to do. Paul said in Romans 7:15 *"For that which I do I allow not: for what I would, that do I not; but what I hate, that do I."*

Imagine the following: Honey, take out the trash. "Great, that's what I really wanted to do." Honey, wash the car. "Wonderful, that's what I feel

Appreciation

like doing." Honey, my shirts need ironing. "Fantastic, I really wanted to do that." Honey, Let's clean out the attic today. "How did you know I've wanted to do that?"

Most things in life followed by our feelings don't get done:

Think about some foods you don't like. Remember President Bush didn't like broccoli. Someone sent him a package to the White House. On television he said, "Watch my lips, I-don't-eat-broccoli."

One of our daughters didn't like liver. Her mother made her eat some anyway. She cut it into small pieces and swallowed it like taking a pull.

Try eating a small amount anyway. You may soon you will learn to like it.

Appreciation

Fake it 'til you make it:

O. H. Mowrer said, "It is easier to act yourself into a better way of feeling than to feel yourself into a better way of acting."

Do we sing because we are happy or are we happy because we sing? The baby wakes up at night and wants the diaper changed. You're tired and don't feel like getting up. You do it anyway. Then you feel better because you acted in a responsible way.

When depressed, get your mind off yourself. Do something good for someone else you will start to feel good. You have to go through the motion to feel the emotion.

Don't wait for the urge:

The odds are good that things will never get done if we wait for the urge. You think the lumberjack has an urge to cut down big trees?

Appreciation

You cannot <u>will</u> a feeling. You can't <u>think</u> yourself out of depression. You can <u>will</u> an <u>action</u> that will control a feeling.

Get up in the morning and shout, "Boy do I feel good." Soon you will feel good. You cannot feel good by thinking it; you have to say it (action).

Peter didn't wait, he took action. John 21 - *"Afterward Jesus appeared again to his disciples, by the Sea of Tiberias. It happened this way: Simon Peter, Thomas (called Didymus), Nathanael from Cana in Galilee, the sons of Zebedee, and two other disciples were together. I'm going out to fish," Simon Peter told them, and they said, 'We'll go with you.' So they went out and got into the boat, but that night they caught nothing. Early in the morning, Jesus stood on the shore, but the disciples did not realize that it was Jesus."*

Appreciation

Listen to your mind:

Do what your mind tells you to do. The attitudes and feeling will follow. God didn't give you a feeler, he gave you a thinker. Our brain (mind) is designed for that purpose. Listening to your feeler will blow your diet, bring failure to your job, distract you from your goals and destroy your marriage. Almost any partner who has left a marriage will say, "He or she makes me feel good. They are thinking with their feeler and not their brain.

Act as if you are happy:

Act like a happy couple whether you feel like it or not. Do something good for your spouse. It has a snowball effect.

Will it work? You don't know 'til you go and see. My five year old granddaughter got to the table

Appreciation

to eat breakfast and she didn't see her favorite cereal. She told her Dad, "I don't like this cereal." He said, "Honey eat it this time and tomorrow we will have your cereal." She said, "I want my cereal now." He said, "We don't have any." She said, "Go down to the store and get some." Her Dad said, "Honey they don't sell your kind of cereal." She replied, "You don't know till you go and see."

Heb. 11:1 – *"Now faith is being sure of what we hope for and certain of what we do not see."*

Heb. 11:32-34 – *"And what more shall I say? I do not have time to tell about Gideon, Barak, Samson, Jephthah, David, Samuel and the prophets, who through faith conquered kingdoms, administered justice, and gained what was promised; who shut the mouths of lions, quenched he fury of the flames, and escaped the edge of the sword; whose weakness was turned to strength; and who became powerful in battle and routed foreign armies."*

Appreciation

Go the extra mile:

In my book, Frank & Mabel, the Governor passed a law that when a military person asked you to carry their back pack you were obligated to carry it one mile. Frank was a farmer and was busy with the spring gardening. He drove a steel peg in front of his house and measured a mile and drove another one. In the course of time Frank because a Christian and learned about going the extra mile. The soldiers had learned that Frank had the two pegs. One day Frank carried the back for a mile and remembered the sermon he heard last Sunday. When they got to the second peg the soldier said, I'll take it now." Frank said; Let me take a while further." When he got to the second mile he gave the pack to the soldier and told him to have a good day.

(Matthew 5: 38-41) "You have heard that it was said, `Eye for eye, and tooth for tooth.' But I tell you, do not resist an evil person. If someone strikes you on the right cheek, turn to him the other

Appreciation

also."

Luther King was a true example of this principle.) "And if someone wants to sue you and take your tunic, let him have your cloak as well. If someone forces you to go one mile, go with him two miles."

Do more than is required. This applies to relationships with other people as well as service to God. A statement I use often is: "A good rule of thumb, do what is required and then some."

CHAPTER FOUR
Use You Compass When You Get Lost

Men have the reputation for refusing to stop and ask for directions. My wife, son and daughter-in-law were traveling with me through West Virginia when the road ended without a sign telling where to go. While stopped and looking at the map a guy on a motorcycle approached us. He said this happens all the time. He instructed us to follow him. We went around some curves that went had to stop and back up and get another start to make the

Appreciation

curve. After a while he stopped and told us to keep going on the same road and it would take us to where we wanted to go. After many miles the road ended again. In the distance we could see that they were building interstate I-77 highway. We cross the field, some ditches and finally got on the new highway. We got to a toll booth. The attendant asked for our ticket. We told him we don't have one. He asked how we got on the highway. I don't think he believed our story.

Life needs a purpose:

Sometimes our purpose is too small. Often we confuse a goal and a purpose. They are not the same. To say, "We I die I want to go to heaven," is not a goal. It is a purpose. A goal must have three parts; what do you want, when do you want it and what will you do to get it.

Appreciation

We are judged by our aims:

Mark 9:34—*"They came to Capernaum. When he was in the house, he asked them, What were you arguing about on the road?" But they kept quiet because on the way they had argued about who was the greatest. Sitting down, Jesus called the Twelve and said, If anyone wants to be first, he must be the very last, and the servant of all."*

Goals are obvious:

What goals do you have for your marriage? Paul Faulkner, in his book, Making Things Right When Things Go Wrong said, "Most couples I counsel with do not have goals. He asked, "What about a will? What about a plan for children's education? What about a plan for living a long happy life together?"

Sign: "Do you know where you are going?"
- God

Appreciation

Drifting through life is life traveling without a compass:

Matt. 6:24 - "No one can serve two masters. Either he will hate the one and love the other, or he will be devoted to the one and despise the other. You cannot serve both God and Money.

1 Kings 8:21 - *"Elijah went before the people and said, How long will you waver between two opinions? If the LORD is God, follow him; but if Baal is God, follow him." But the people said nothing."*

Acts 2:40 – *"And with many other words did he testify and exhort, saying, Save yourselves from this untoward generation."* (KJ) *Save yourselves from this crooked generation.* (ASV) *Save yourselves from this corrupt generation.* '(NIV)

Goals are important:

Dr. Viktor Frankl, In Man's Search For

Appreciation

Meaning. "Prisoners in the German concentration camps could endure almost anything as long as they had a purpose."

2 Cor. 13:11 - *"Finally, brothers, good-by. Aim for perfection, listen to my appeal, be of one mind, live in peace. And the God of love and peace will be with you."*

Goals come with power:

Ralph Waldo Emerson said, "You can run the Gulf Stream through a drinking straw, if you place the straw parallel to the flow of the stream."

Our lives need to be lined up with God's purpose. 40 Days of Purpose, has been one of my favorite books.

God's power flows through us if we are aligned with His purpose. Goals help us to focus on our lives by concentrating on a specific direction.

Eph. 3:20 - *"Now to him who is able to do*

Appreciation

immeasurably more than all we ask or imagine, according to his power that is at work within us,"

Concentrate your efforts:

Try devoting your attention to 101 things; soon you will lack the ability to focus on anything. Some tasks do not need concentration. A writer was told, "If you are going to write a book, your grass will have to grow long."

Develop a gleam in your eye:

God put a gleam in the eyes of Moses. He accomplished many great things.

Homeless to Harvard was a remarkable story. Great things can be accomplished by ordinary people when they have a gleam in their eyes.

The single-eye approach:

1 Peter 1:13 – *"Therefore, prepare your*

Appreciation

minds for action; be self-controlled; set your hope fully on the grace to be given you when Jesus Christ is revealed."

Matt. 6:22 - *The eye is the lamp of the body. If your eyes are good, your whole body will be full of light. But if your eyes are bad, your whole body will be full of darkness. If then the light within you is darkness, how great is that darkness!"*

When we try to serve two masters we are cross-eyed. We have to be disciplined to reach our goals.

Goals bring harmony to our lives:

Decide where you are going. Whether you are going to New York or to Miami makes a big difference in the direction you start in.

Proverbs 2: 1-6— *"My son, if you accept my words and store up my commands within you, turning your ear to wisdom and applying your heart to*

Appreciation

understanding, and if you call out for insight and cry aloud for understanding, and if you look for it as for silver and search for it as for hidden treasure, then you will understand the fear of the LORD and find the knowledge of God. For the LORD gives wisdom, and from his mouth come knowledge and understanding."

Use the talents God has given you:

Imagine a 55 gallon barrel and a thimble on the floor. Some people have a barrel full of talents. Others have a tiny thimble full.

Use what you have to gain more. Matthew 25:14 *"For the kingdom of heaven is as a man travelling into a far country, who called his own servants, and delivered unto them his goods. 15 And unto one he gave five talents, to another two, and to another one; to every man according to his several ability; and straightway took his journey. 16 Then he that had received the five talents*

Appreciation

went and traded with the same, and made them other five talents. 17 And likewise he that had received two, he also gained other two. 18 But he that had received one went and digged in the earth, and hid his lord's money. 19 After a long time the lord of those servants cometh, and reckoneth with them. 20 And so he that had received five talents came and brought other five talents, saying, Lord, thou deliveredst unto me five talents: behold, I have gained beside them five talents more. 21 His lord said unto him, Well done, thou good and faithful servant: thou hast been faithful over a few things, I will make thee ruler over many things: enter thou into the joy of thy lord. 22 He also that had received two talents came and said, Lord, thou deliveredst unto me two talents: behold, I have gained two other talents beside them. 23 His lord said unto him, Well done, good and faithful servant; thou hast been faithful over a few things, I will make thee ruler over many things: enter thou into the joy of thy lord. 24 Then he which had received the one

Appreciation

talent came and said, Lord, I knew thee that thou art an hard man, reaping where thou hast not sown, and gathering where thou hast not strawed: 25 And I was afraid, and went and hid thy talent in the earth: lo, there thou hast that is thine. 26 His lord answered and said unto him, Thou wicked and slothful servant, thou knewest that I reap where I sowed not, and gather where I have not strawed: 27 Thou oughtest therefore to have put my money to the exchangers, and then at my coming I should have received mine own with usury. 28 Take therefore the talent from him, and give it unto him which hath ten talents. 29 For unto every one that hath shall be given, and he shall have abundance: but from him that hath not shall be taken away even that which he hath."

"It is finished:"

One of the things Jesus said on the cross, it is finished. Were all the people healed? No. Were all

Appreciation

the hungry fed? No. But Jesus finished what God sent Him here to do.

When it's your time to go . . .

- There may be weeds in the garden.

- The lawn may need to be mowed.

- There may be dirty dishes in the sink.

Do like Jesus did, "I came to be about my Father's business."

CHAPTER FIVE
Don't Kill Today With Yesterday

Philippians 3:13-14 - *"Brothers, I do not consider myself yet to have taken hold of it. But one thing I do: Forgetting what is behind and straining toward what is ahead, I press on toward the goal to win the prize for which God has called me heavenward in Christ Jesus."*

Guilt causes stress. We have to find ways to overcome guilt.

Ralph Waldo Emerson wrote a letter to his daughter; *"Finish every day and be done with it.*

Appreciation

You have done what you could. Some blunders and absurdities no doubt crept in; but get rid of them and forget them as soon as you can. Tomorrow is a new day, and you should never encumber potentialities and invitations with the dread of the past. You should not waste a moment of today on the rottenness of yesterday.

Maybe we made blunders and failures of our relationships. How can we get over our failures? How can we forgive ourselves and begin again? Many young ladies get pregnant out of wedlock and the parents force them to have an abortion. The trauma sometimes lingers for a lifetime.

Sometimes a relative dies and we remember the bad times. How can I forgive myself for not spending more time with my children when they were younger? You can't relive the past. Learn to accept the things you cannot change. Forgive yourself of the mistakes you made. Complete forgiveness is necessary.

Appreciation

Come to Grips with the past:

Some people handle this better than others. They stumble and fail, then try again. Talk to yourself about the past, *"Self, the past is gone, there is nothing you can do to change it. So why waste your strength wrestling with the past when you need all your strength to face the issues of today."*

In one of Shakespeare's plays, one line is, *"I've shut the door on yesterday and I've thrown away the key…"*

Our shoulders are not strong enough to carry all the loads of the past. Only God is strong enough.

Deuteronomy 33:25 – *"Thy bars shall be iron and brass; And as thy days, so shall thy strength be."* (ASV)

God promises strength for today. Matthew 6:34 – *"Be not therefore anxious for the morrow: for the morrow will be anxious for itself. Sufficient*

Appreciation

unto the day is the evil thereof."

Yesterday is a cancelled check, tomorrow is a promissory note, and today is the only legal tender we have.

Sybil Partridge said, *"So for tomorrow and its needs, I do not pray; but keep me, guide me, love me Lord, just for today."*

One day at a time:

"I can make it, one day at a time."

God's power comes as we yield to His will. Psalms 118:24 – *"This is the day the LORD has made; let us rejoice and be glad in it."*

Don't look back at past mistakes. Luke 9:62 – *"Jesus replied, No one who puts his hand to the plow and looks back is fit for service in the kingdom of God."*

A newly hired farm hand was in the field plowing with the farmer. He stopped and asked,

Appreciation

"Why my rows crooked and yours are are straight." The farmer replied, "When I start a new row I pick a tree or post straight ahead and I keep my eye on it." When you plow you keep looking back to see how you are doing. Forget what's been done and look ahead.

Paul said, *"Forgetting what is behind and straining toward what is ahead..."*

Don't let yesterday's mistakes paralyze your efforts today.

Problems with living in the past:

- Technically it's impossible.

- We are locked into a time capsule that drags us through life.

- We don't function well thinking about the past.

We learn in sports to keep our eyes on the goal. Someone said God put our eyes in front so we

Appreciation

could go forward. We don't do well looking over our shoulder. Past memories are flawed.

Remember the good ol' days?

- You could get a Pepsi for five cents.

- You could get a dollar's worth of gas and get free services.

We forget about . . . Walking down a path to the toilet. Having to heat water in a tin tub to take a bath. Going outside and pump your water. Start the car with a crank. Your clothes washer was on the back porch with two rollers. Walk to the country store to use the telephone. Listening to the Grand Old Opry on the radio. The 60 HP fords that wouldn't climb a hill without changing gears. The good old days are faulty.

Sometimes our computer disk gets full. When we overload our mind and body things begin to malfunction.

Appreciation

Guilt:

Two kinds of guilt. One comes from the judgments and suggestion one makes. Childhood memories of something you thought was bad. Something that violates your conscience.

Moral code. The other comes from Divine judgment. Violating God's laws; murder, lying, stealing, adultery ... Romans 3:23 - *"for all have sinned and fall short of the glory of God,"*

The grace of God takes all guilt away. When we became Christians our sins were washed away. When we sin, if we are penitent, Christ blood continues to wash our sins away. It's like waking in a river with a muddy bottom . . . The mud is sin and the river is the blood of Christ. Every time you lift your muddy foot, the blood of Christ washes away the sin.

The price of guilt:

David sins. Psalms 32: 3-5 - *"When I kept*

Appreciation

silent, my bones wasted away through my groaning all day long. For day and night your hand was heavy upon me; my strength was sapped as in the heat of summer. Then I acknowledged my sin to you and did not cover up my iniquity. I said, 'I will confess my transgressions to the LORD'- and you forgave the guilt of my sin."

Proverbs 28:13 - *"He who conceals his sins does not prosper, but whoever confesses and renounces them finds mercy."*

How Guilt works. When you violate your moral values, you create a psychological debt. The debt has to be paid back, one way or another. The conscience is the debt collector. It makes you sick, headaches, stomach pains, etc. until the debt is paid.

The American Indians picture the conscience as a triangle, three cornered stone. When you violate your values the stone turns. If you continue the stone will round off the sharp edges.

Appreciation

Vain attempts to pay off the debts:

- The immature attempt; I'll do enough good things to make it even. We can't earn our way to Heaven.
- The depressive attempt; I'll grieve over it. They become depressed.
- The neurotic attempt; I'll get sick. When you don't want to go to church, you can get sick.
- The criminal attempt; don't blame me . . . It's the neighborhood I live in.
- It runs in my family. It's in my blood, I can't help it. The abuser says, my parents were abusive, don't blame me.

Passing the buck is how it all started. Adam and Eve in the Garden of Eden, "He made me do it."

Appreciation

How do I forgive myself?

Debts we owe God cannot be paid by our efforts. Only God can forgive our sins. God's love paid our debts.

Song:

He paid a debt He did not owe; I owed a debt I could not pay; I needed someone to wash my sins away, And now I sing a brand new song—amazing grace! Christ Jesus paid the debt that I could never pay.

CHAPTER SIX

Cut Your Line When It Gets Tangled

What do you do when your fishing line gets tangled?

Years ago carpenters would spend a tremendous amount of time trying to get a knot of a line. Now we have learned that it's best to cut the line and pull out some new line. The same goes for fishermen.

Some say, "I'm going to straighten this thing out before I go any further." Some instances it may

Appreciation

be worth the hassle.

The thrill of the drill:

Some people like to drill for the thrill of it. They love bringing ringing up resentments. You may have temporarily put this in your subconscious. This happens when one is angry, they bring up resentments.

What is resentment?

Latin word *"resento"* meaning to re-feel. That painful memory is brought to the surface. You re-feel the hurt, and agony you are trying to forget. The best way the handle resentment is "cut the line." Randy Travis, a country singer, calls it "Digging up bones."

Take Paul's advice, Put the past behind you and strive for what is ahead. Don't waste a lot of energy and hurt feelings with resentment. Did you

Appreciation

ever live on a farm? If you did, you know what a slop bucket is. Bringing up resentments is pouring the slop over their heads.

Christ took the initiative:

How can you forgive someone who hates you? That's what Jesus did in the cross. Pilate knew he was not guilty. He thought, I'll scourge Him and let Him go. The angry mob didn't accept that. Luke 23:34 - *Jesus said, Father, forgive them, for they do not know what they are doing."*Jesus took the initiative to forgive, even though they didn't ask.

Acts 2:23,37,38 – *"This man was handed over to you by God's set purpose and foreknowledge; and you, with the help of wicked men, put him to death by nailing him to the cross. When the people heard this, they were cut to the heart and said to Peter and the other apostles, Brothers, what shall we do?" Peter replied, Repent and be*

Appreciation

baptized, every one of you, in the name of Jesus Christ for the forgiveness of your sins. And you will receive the gift of the Holy Spirit."

Romans 5:6-8 – *"You see, at just the right time, when we were still powerless, Christ died for the ungodly. Very rarely will anyone die for a righteous man, though for a good man someone might possibly dare to die. But God demonstrates his own love for us in this: While we were still sinners, Christ died for us."*

When people mistreat us and resentments arise, forgive them. We need to take the initiative to forgive them. Taking the initiative is cutting the line. Some play the game, I'll forgive you if you will forgive me.

1 Peter 2:21-24 – *"To this you were called, because Christ suffered for you, leaving you an example, that you should follow in his steps. He committed no sin, and no deceit was found in his mouth." When they hurled their insults at him, he*

Appreciation

did not retaliate; when he suffered, he made no threats. Instead, he entrusted himself to him who judges justly. He himself bore our sins in his body on the tree, so that we might die to sins and live for righteousness; by his wounds you have been healed."

Peter was preparing us for what he said later. 1 Peter 3:8-9 – *"Finally, all of you, live in harmony with one another; be sympathetic, love as brothers, be compassionate and humble. Do not repay evil with evil or insult with insult, but with blessing, because to this you were called so that you may inherit a blessing."*

Eph. 3:32 – *"Be kind and compassionate to one another, forgiving each other, just as in Christ God forgave you."*

The case of Job:

He cried out in bitterness, God I want an um-

Appreciation

pire between us. Same as saying God you're giving me a raw deal; I want someone to settle this dispute who will be fairly.

God dealt graciously with Job in His own time, and he will with us. Maybe your rights have been violated …

You've been passed up for a promotion,

Your spouse has treated you wrong,

A thief stole your belongings,

You were abused as a child,

Your stock broker chose the wrong stock and you lost money,

Someone told lies on you…

Take the initiative to forgive them.

Appreciation

How do you overcome resentments?

All the things we read. "Love you enemies." "Do good to those who hate you." "Overcome evil with good." The big job is how do we translate these principles into our daily behavior?

Become vulnerable. When you respond with love to evil, you expose yourself to cruel treatment, but it's worth it.

Learn to accept rejection. Some who have learned what it's like to get rejections:

- Writers . . .

- Artist . . .

- Inventors . . .

Be forgiving. Forget what is behind, "Cut the line." We learn to live with scars. When you cut yourself even after it heals there may still be a scar.

Keep on blessing others. I know that's hard to do. But, being a Christian is tough.

Appreciation

People usually ask three questions:

How do I get even? You don't. Romans 12:17-19 - *"Do not repay anyone evil for evil. Be careful to do what is right in the eyes of everybody. If it is possible, as far as it depends on you, live at peace with everyone. Do not take revenge, my friends, but leave room for God's wrath, for it is written: It is mine to avenge; I will repay,"says the Lord."*

How do I get rid of the pain? You may have to live with some of it. Like when the sore heals, you sometimes still have a scar and sometime the pain will linger.

Try turning your pain into strength. Paul had a thorn in the flesh. Look for the good in the situation.

How do I get rid of the memories? Turn them over to God. Remember the good times. The story of Joseph in the book of Genesis shows that we can forgive others when they have wronged us.

Chapter - 7

Stay Cool, Even When You're Hot

Sports ... Hockey, basketball ... Remember the basketball game when the coach threw a chair into the audience.

What makes you angry?

Definition: **Anger:** means an intense emotional state induced by displeasure. The most general term, names the reaction but in itself conveys nothing about intensity. **Rage:** suggests loss of self-control from violence of emotion. **Fury:** destructive rage that can verge on madness. **Indignation:** stresses righteous anger at what one considers unfair, mean, or shameful. **Wrath:** is likely to suggest a desire or intent to revenge or punish.

Anger within itself is not wrong.

Exodus 4: 14 – *"Then the LORD's anger burned against Moses."*

Matt. 21:12 - *"And Jesus entered into the temple of God, and cast out all them that sold and bought in the temple, and overthrew the tables of the money-changers, and the seats of them that sold the doves;"*

Eph. 4:26 - *"Be ye angry, and sin not:"* (KJV) *"In your anger do not sin"* (NIV)

Titus 1:7 - *"In your anger do not sin"* (NIV) *"Not soon anger"* (KJV)

Appreciation

Get a handle on anger:

A quick flash of anger destroys our best intentions.

The story of Cain; Gen. 4:1-8 – *"Adam lay with his wife Eve, and she became pregnant and gave birth to Cain. She said, With the help of the LORD I have brought forth a man." 2. Later she gave birth to his brother Abel. Now Abel kept flocks, and Cain worked the soil. 3. In the course of time Cain brought some of the fruits of the soil as an offering to the LORD. 4. But Abel brought fat portions from some of the firstborn of his flock. The LORD looked with favor on Abel and his offering, 5. but on Cain and his offering he did not look with favor. So Cain was very angry, and his face was downcast. 6. Then the LORD said to Cain, Why are you angry? Why is your face downcast? 7. If you do what is right, will you not be accepted? But if you do not do what is right, sin is crouching at your door; it desires to have you, but you must*

Appreciation

master it." 8. Now Cain said to his brother Abel, Let's go out to the field." And while they were in the field, Cain attacked his brother Abel and killed him."

What we need to learn is, when we lose control, we are totally out of control.

Anger is supercharged.

- Has incredible effects on the body.

- Constricts the blood vessels.

- Dilates the eyes. "I'm so angry I can't see straight."

- Anger causes stress.

A Harvard professor said, "Anger causes stress and stress causes physical illness." Dr. Carl Simonton says one of the root causes of cancer is to hold in resentments. He says there is a remarkable remission of the disease when the patient learns to deal with anger.

Duke University found that anger produces 2

Appreciation

to 5 times more deaths than high blood pressure and smoking.

Anger prevents problem solving. It shuts down an open mind. Two people may get along fine until one gets angry. Research shows that 68% of problems are not solved because of anger.

Anger distorts the truth. People make overblown statements like: "You're always late." "You never shut the door."

Anger is addictive. People that get angry often decide it's the only way to handle problems. Someone said, give a speech when you are angry, it will be the best speak you'll ever regret.

What causes anger?

Your rights are violated: You feel passed up for a promotion. A car comes around you and cuts you off. One of my daughters used to say, "Don't get mad consider the circumstances… A car fails to

Appreciation

yield the right of way; you get steamed because you feel you have been treated unfairly.

Unrealistic expectations: You feel, they did it this way at my last company, why can't they do it here. This is the way it was with my last marriage, why can't it be that way now.

Sam Walton used to have a plague in his office: "Don't treat people as they really are, treat them as you want them to become."

Three ways anger:

Vent the anger. Psychologists say, "Get it out." Do it in a positive way. Remember the dog in the cartoons - He gets hurt and runs over the hill where no one can hear him and lets it out. Clean the garage. A teenage girl used to say, I know when my Daddy is mad, he starts cleaning the garage." Work in the garden. Cut the grass

Proverbs 22:24 – *"Make no friendship with a*

Appreciation

man that is given to anger; And with a wrathful man thou shalt not go:"

Proverbs 29:11: - *"A fool gives full vent to his anger, but a wise man keeps himself under control."*

Proverbs 15:18 – *"A hot-tempered man stirs up dissension, but a patient man calms a quarrel."*

Stuff it. People react to anger is different ways. They clam up and keep it inside. This is like taking poison. It causes stomach aches and headaches. If you don't let it out and you don't hold it in, what do you do?

Prevent it. An ounce of prevention is worth a pound of cure. Or, an ounce of prevention is worth a pound of apologies.

James 1:19-21 - *"My dear brothers, take note of this: Everyone should be quick to listen, slow to speak and slow to become angry, 20. for man's anger does not bring about the righteous life*

Appreciation

that God desires. 21. Therefore, get rid of all moral filth and the evil that is so prevalent and humbly accept the word planted in you, which can save you."

How do you know when someone is angry?

They change their tone of voice. The voice gets higher. They emphasis every word, "I... told... you... before... "

Sometime it's silence. It is easy to tell when some people are angry. They turn red in the face.

How should we respond to signs of anger?

Do it as quick as you can. If you see someone is getting out of control. I'm sorry, I must have messed up. I guess I said the wrong thing. Where did I go wrong? That's not what I meant to say.

Appreciation

Did anyone ever make you angry on purpose?

You know the "hot buttons" of people close to you. Paul knew the hot buttons of certain people. Acts 23:6-7 – *"Then Paul, knowing that some of them were Sadducees and the others Pharisees, called out in the Sanhedrin, My brothers, I am a Pharisee, the son of a Pharisee. I stand on trial because of my hope in the resurrection of the dead." 7. When he said this, a dispute broke out between the Pharisees and the Sadducees, ..."*

Four steps to processing anger:

Give permission to be angry:

Anger within itself is neither right nor wrong.

Sometimes anger is good. When a criminal gets out of jail on a technicality, the policeman gets angry. When we see people abused, it makes us an-

Appreciation

gry.

Mark 3:1-5 - *"Another time he went into the synagogue, and a man with a shriveled hand was there. 2. Some of them were looking for a reason to accuse Jesus, so they watched him closely to see if he would heal him on the Sabbath. 3. Jesus said to the man with the shriveled hand, Stand up in front of everyone." 4. Then Jesus asked them, Which is lawful on the Sabbath: to do good or to do evil, to save life or to kill?"But they remained silent. 5. He looked around at them in anger and, deeply distressed at their stubborn hearts, said to the man, Stretch out your hand."He stretched it out, and his hand was completely restored."*

Report feelings of anger. People don't always know when you are angry. Tell them, "You hurt my feelings." Don't report anger in front of other people.

Make a "no hurt contract." Tell your family and friends, if I hurt your feeling let me know. Will

Appreciation

you help me through my anger? Tell them I have a problem, will you help me? Agree to disagree.

CHAPTER EIGHT

Make Your Relationships Right

2 Cor. 5:18-20 - *"Therefore, if anyone is in Christ, he is a new creation; the old has gone, the new has come! 18. All this is from God, who reconciled us to himself through Christ and gave us the ministry of reconciliation: 19. that God was reconciling the world to himself in Christ, not counting men's sins against them. And he has committed to us the message of reconciliation. 20. We are therefore Christ's ambassadors, as though God were making his appeal through us. We implore you on Christ's*

Appreciation

behalf: Be reconciled to God."

Definition of reconcile: To restore to friendship or harmony; to cause to submit to or accept something unpleasant; to settle or resolve.

God made everything perfect. Man sinned. Because of man's sins, he gave a way of reconciliation. As reconciled Christians, we have the responsibility of reconciling others.

We are ambassadors. God works through us. God does not save anyone automatically. He uses people as ambassadors. Ambassadors have a specific responsibility. A survey shows everyone has a positive influence over an average of six people.

Just as God uses us to reconcile sinners to Him, he uses us to reconcile relationships in our lives.

We cannot leave it alone and let it will go away. Sometimes relationships are broken suddenly.

Appreciation

People need people:

There are no happy hermits. God intended for us to be social beings. We depend on one another. Song, "People who need people are the luckiest people in the world. People who withdraw from people have problems.

A Jewish Rabbi said, "Anyone who goes too far alone goes mad." A student said, "I feel like one page in a thousand page book, which no one reads."

What if someone offered you the car of your choice, the house of your choice and all the money you would ever need. There is just one catch. You will have to live on a deserted island alone the rest of your life.

We buy things to impress other people. Look what I just bought … Let me tell you about … If we had everything, we would still need someone to share it with.

Appreciation

Who are people?

They are a spirit inside a body. When we get to Heaven, how will we recognize people? Our fleshly bodies will be gone. We will recognize one another's spirits. Things don't honor us, people do.

Why should we love and get along with people?

It is good medicine. "It did her good to see her friend." One professor said, "By people we are broken and by people we are put together again." In a study at Berkley, "People with weak social ties have 2 – 5 times higher death rate than people with strong social ties." That's the purpose in meeting often. Somebody said you got to have six friends, that's how many you need to put you in the ground. It's good medicine to do go for someone.

One reason to love and get along with people is you feel better about yourself. When someone

Appreciation

gives you a compliment it makes you feel better. Someone tells you that you are important it makes you feel better. Everyone needs a pat on the back.

People are a good investment. When our parents or friends get old we sometimes ask, "What do you need?"

Joe DiMaggio said Marilyn Monroe had everything to live with, but nothing to live for."

Matt. 6:19-21 - *Do not store up for yourselves treasures on earth, where moth and rust destroy, and where thieves break in and steal. 20. But store up for yourselves treasures in heaven, where moth and rust do not destroy, and where thieves do not break in and steal. 21. For where your treasure is, there your heart will be also."*

We need to invest in good relationships. It does mean money. We can invest in time and a good attitude. Sometimes parents fail to invest in relationships with their children.

CHAPTER NINE

Go First Anyway

Gal. 5:13 – *"You, my brothers, were called to be free. But do not use your freedom to indulge the sinful nature ; rather, serve one another in love."*

They were living under a law of bondage. Now they are under a law of liberty. Paul is saying, don't use the freedom you have as an opportunity to sin, or be partakers of the works of the flesh.

Gal. 5:19 - 21 - *Now the works (deeds, acts and things done) of the flesh (earthly nature of*

Appreciation

man) are manifest, (to make visible) which are these; Adultery, fornication (a form of impurity either before or after marriage), uncleanness (sexual Impurity), lasciviousness (to lust), Idolatry (to worship idols), witchcraft (the use of magic to deceive others), hatred (ill will), variance (being disagreeable), emulations (an unfriendly feeling), wrath (outburst of anger), strife (selfishness), seditions (divisions), heresies (division caused by opinions), envyings (quarrels that lead to grudges), murders (to take another's life), drunkenness (intoxication), revellings (drunken parties), and such like in case I left something out, includes but does not exclude:

Rather use your freedom to serve one another.

Gal. 6:10 – *"As we have therefore opportunity, let us do good (serve) unto all men, especially unto them who are of the household of faith."*

Appreciation

Example of serving others:

The story of the Good Samaritan is an excellent example of serving others. A man was robbed and left for dead on the side of the road. A priest and a Levite came by and went on their way without offering to help him. A Samaritan, one of another race, came by and saw that he was hurt. He put him on his donkey and carried him to the Inn. He told the Inn keeper, "Give him what he needs. Here is some money and if it is not enough I will pay you when I come back."

Servants are also called slaves or ministers.

How do we serve others?

Romans 12:5 - *"so in Christ we who are many form one body, and each member belongs to all the others."*

Verse 10 – *"Be devoted to one another in brotherly love. Honor one another above your-*

Appreciation

selves."

Verse 14 – *"Bless those who persecute you; bless and do not curse."*

Verse 15 – *"Rejoice with those who rejoice; mourn with those who mourn."*

Verse 16 – *"Live in harmony with one another. Do not be proud, but be willing to associate with people of low position. Do not be conceited."*

Verses 19-21 – *"Do not take revenge, my friends, but leave room for God's wrath, for it is written: It is mine to avenge; I will repay," says the Lord. On the contrary: If your enemy is hungry, feed him; if he is thirsty, give him something to drink. In doing this, you will heap burning coals on his head."Do not be overcome by evil, but overcome evil with good."*

Learn to see the potential in others:

Jesus saw potential in the woman at the well.

Appreciation

He helped her to see her mistakes. When she learned about Jesus, she convinced the whole town to come and hear Jesus. She evangelized an entire city by herself.

What do good servants do?

First things they don't do. They don't care who gets the credit. They don't get their feeling hurt when their name is not in the bulletin. They don't keep score. They don't play the martyr. They don't feel sorry for themselves.

Things a good servant will do. They listen louder. Listening is a talent. They are transparent. Their life is an open book.

ames 5:16 – *"Therefore confess your sins to each other and pray for each other so that you may be healed. The prayer of a righteous man is powerful and effective. "*

A servant is sensitive. They care about how

Appreciation

others feel. Servants learn to look down the ladder. We live in a ladder climbing society.

Learn to go first, even if it is not your turn:

To make our relationships right we have to take the initiative to say, "I'm sorry, please forgive me."

Matt. 20:25-28 – *"Jesus called them together and said, You know that the rulers of the Gentiles lord it over them, and their high officials exercise authority over them. Not so with you. Instead, whoever wants to become great among you must be your servant, and whoever wants to be first must be your slave-- just as the Son of Man did not come to be served, but to serve, and to give his life as a ransom for many."*

Appreciation

Anyway

People are unreasonable, illogical and self-centered. **Love them anyway.** If you do good, people will accuse you of selfish ulterior motives. **Do good anyway.** If you are successful, you'll win false friends and true enemies. **Succeed anyway.** Honesty and frankness make you vulnerable. **Be honest and frank anyway.** The good you do today will be forgotten tomorrow. **Do good anyway.** The biggest problem with the biggest ideas it can be shot down by the smallest people with small minds. **Think big anyway.** People favor underdogs, but follow top dogs. **Fight for some underdogs anyway.** What you spend years building may be destroyed overnight. **Build anyway.** Give the world the best you've got, and you'll get kicked in the teeth. **Give the world the best you've got anyway.**

- Reader's Digest

CHAPTER TEN

Live Young

Even When You Are Old

Choose to die young, no matter how old you are:

When the weather turned cold, a farmer got his rifle and went to the hog pen to kill his hog. When he got there he remembered he killed the hog last year.

    Gen. Douglas MacArthur - "Youth is not a time of life; it is a state of mind. You are as young

Appreciation

as your faith, as old as your doubts, as young as your self confidence, as old as your fears, as young as your hopes, as old as your despairs."

Someone said, "The lines on my face are service stripes of yesterday, which have equipped me to do a better job of helping people today."

How do you know when you are getting old?

- You can tell you are getting older when everything hurts, and what doesn't hurt doesn't work.
- Your little black book only contains names that end with "MD."
- Your children began to look middle aged.
- You look forward to a dull evening.
- You sit in a rocking chair and can't get in going.

Appreciation

- Your knees buckle, but your belt won't.
- Your pace maker makes the garage door open when a pretty girl goes by.
- You sink your teeth into a steak and they stay there.
- The little gray-haired lady that helps you across the street is your wife.

Never stop:

There are physical limitations with age. Someone said, "I can't jump as high, but I can stay up longer." When I stoop down to pick up something, I look around to see what else is down there I can get. It's not the number of candles on the cake, it's your attitude.

Keep moving and climbing until you have fulfilled you mission on earth. Jesus died at age 33 1/2. His mission was fulfilled. Moses died at a

Appreciation

young age of 120. Caleb was still strong at age 85. Joshua 14:10-12 - *Now then, just as the LORD promised, he has kept me alive for forty-five years since the time he said this to Moses, while Israel moved about in the desert. So here I am today, eighty-five years old! I am still as strong today as the day Moses sent me out; I'm just as vigorous to go out to battle now as I was then. Now give me this hill country that the LORD promised me that day. You yourself heard then that the Anakites were there and their cities were large and fortified, but, the LORD helping me, I will drive them out just as he said."*

Grandma Moses started painting when she was 80.

Colonel Sanders was retired when he started his chicken business.

I personally laid bricks until I was 75.

Appreciation

Growing and changing:

Winners in life are always changing. They are always looking for a better way. Losers just do their job, or enough to get by. They don't get promoted. They're not much fun to be around. Where there is life there is growth. Anything not growing is dying.

Our bodies change. We don't look the same as we did 20 years ago. If you think you do, look in a mirror. Our minds are changing. Our values are changing. Our personalities and characteristics are changing. Our friendships are changing. We have to keep in step with the parade or life will pass us by.

Making adjustments:

If you want to grow in a personal relationship, the only person that can do it is you. You may have to make adjustments. I know many who waited for their children to get through high school and they divorced. They didn't make adjustments in

Appreciation

their lives.

Some couples feel guilty when they are told they need to make adjustments in their lives to learn to live together. They don't want to admit they are doing something wrong. A good relationship sometimes takes hard work. You have to work hard to learn to play tennis, ice skate or learn to swim.

The growth concept:

The Bible doesn't tell us anything about Jesus' life from 12 to 30.

Luke 2:52 – *"And Jesus grew in wisdom and stature, and in favor with God and men."*

Christians are to grow spiritually. John 15:8 – *"This is to my Father's glory, that you bear much fruit, showing yourselves to be my disciples."*

If you want a bushel of grapes tomorrow, you have to start today. 2 Peter 1:5-*8* - *"For this*

Appreciation

very reason, make every effort to add to your faith goodness; and to goodness, knowledge; and to knowledge, self-control; and to self-control, perseverance; and to perseverance, godliness; and to godliness, brotherly kindness; and to brotherly kindness, love. For if you possess these qualities in increasing measure, they will keep you from being ineffective and unproductive in your knowledge of our Lord Jesus Christ."

Psalms 92:12-14 – *"The righteous will flourish like a palm tree, they will grow like a cedar of Lebanon; planted in the house of the LORD, they will flourish in the courts of our God. They will still bear fruit in old age, they will stay fresh and green, proclaiming, The LORD is upright; he is my Rock, and there is no wickedness in him."*

2 Cor. 4:16 - *"Therefore we do not lose heart. Though outwardly we are wasting away, yet inwardly we are being renewed day by day. For our light and momentary troubles are achieving for us*

Appreciation

an eternal glory that far outweighs them all. So we fix our eyes not on what is seen, but on what is unseen. For what is seen is temporary, but what is unseen is eternal. 16:1 - Now we know that if the earthly tent we live in is destroyed, we have a building from God, an eternal house in heaven, not built by human hands."

Physical growth:

Jack LaLanne – at the writing is 88 years old. At age 70 he pulled 70 boats containing 70 people for one mile across the Long Beach Harbor. He did it with a rope in his teeth. His arms were handcuffed.

Noah lived 400 years. He hewed cypress lumber to build the ark.

George Burns and W. Clement Stone live to a ripe old age.

Appreciation

Mental growth:

Tillit S. Teddlie wrote dozens of songs. He was still writing songs at 101.

Lev. 19:32 - *"Rise in the presence of the aged, show respect for the elderly and revere your God. I am the LORD. "*

Will power:

People die young because of a lack of will power. They are tired and want to give up. This book makes more than 80 books I have written. I am 78 years old and have no plans to stop.

The best is yet to be:

Old age can be beautiful. There are peaks and valleys. When death comes, it's just the beginning of eternal life.

All of Raymond E. Smith's book may be viewed at: www.amazon.com/author/papasbooks

www.ingramcontent.com/pod-product-compliance
Lightning Source LLC
Chambersburg PA
CBHW051709170526
45167CB00002B/596